From where will we get the money?

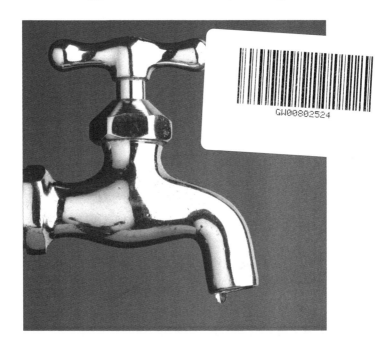

UISCE

*"Reduced To A Trickle..."**

*When Phil Hogan, the Government Minister who introduced the water levy, was asked: what will happen to people who do not pay the water levy - He arrogantly and provocatively replied that: their water supply would be: *"reduced to a trickle"*

All rights reserved. No part of this publication may be copied, reproduced, stored in a retrieval system, or transmitted, in any form or by any means, without the prior permission of the authors, nor be otherwise circulated in any form of binding or cover other than that in which it is published and without a similar condition including this condition being imposed on the subsequent purchaser.

© Pat Keogh 2016

ISBN: 978-1-910179-92-5

Printed and Bound in Ireland by eprint.ie

Acknowledgements

I express gratitude to Tara
for proof-reading this book and offering
suggestions on modifications.
Thanks too, to Páraic for his comments,
and locating Celtic water valve covers

I dedicate this book to the courageous and dedicated men
and women who protested consistently against austerity
and an unjust water levy and particularly the five who
were sent to jail: Derek Byrne, Bernie Hughes, Michael
Batty, Paul Moore and Damien O Neill

also

To the brave men and women of the 1916 Easter Rising,
who fought and died so that we might have a better
Ireland

and

In memory of Frank Doyle
Killarney

Contents

Preface 1

Introduction 3

For Richer. for Poorer - All the Days of Your Life 11

Uisce Faoi Thalamh 29

Irish Water – The Quango 42

Water Conservation 56

Ireland's Debt The European Vision 62

Where Will We Get The Money From? 71

A Lateral Thinking Approach 84

Forestry - Fishing - Tourism 104

Conclusion 109

From where will we get the money?

Preface

Ireland has a bright future. We know that. But we do not want to hear this from our President or Taoiseach. Hearing it from those sources is patronizing and condescending, particularly, since it is difficult to see what they are doing to bring about that bright new future and they live in luxury themselves. Our Taoiseach and his government ministers seem to depend on external circumstances to improve our economy and then they claim the dividend.

A collapse of our political system as swift and dramatic as the collapse of our economy would perhaps create a bright new future for Ireland. The current political system is archaic. It lacks creativity. It is stagnant. It is conservative. It is deep rooted in conniving, grants secured through the political system and jobs for the boys. It is stifling, polluted and regressive.

Ireland is a country rich in basic natural human necessities. It has a small population. It has abundance of fresh, unpolluted air. It has an unparalleled and unequalled amount of fresh clean water. It has a very high rainfall. It has an abundant supply of fish; river, lake and sea fish of numerous varieties. It has excellent grass and tillage land. It has thousands of acres of land suitable for forestry.

Ireland has potentially one of the most magnificent tourist industries in the world, with ancient archaeological and historical sites depicting our heritage, history and the history of humankind. Ireland has numerous natural, beautiful landscapes, mountain views and seascapes. It is a peaceful country.

It has abundance of wealth. That wealth ought to be distributed more equally. It is a land of milk and honey.

It has been said that: *If The Netherlands owned Ireland it could feed all the people in the world, but, if Ireland owned The Netherlands that country would be drowned long ago.*

Introduction

Ireland is one of the wealthiest countries in the world. But the riches are in the hands of the few. Many in the lower bracket of income are living little better than people in the third world, thousands are homeless and many children go hungry every day because their parents cannot afford to feed them. In contrast there are many millionaires living in total luxury in mansions, many of our leading politicians and ministers rub shoulders with the super-rich. The president lives in a mansion while outside the gates homeless Lazarus survives, seeking the crumbs that fall from the rich man's table.

Charging people for water, in a water-rich country, is the last straw for many people who are at the pin of their collar to pay for essentials; food, clothing, health, education and paying essential bills; electricity, gas, insurance, phone, transport, rent or mortgage with constant home repairs, essential replacements of electrical appliances and other everyday home furniture, equipment and utensils. There is also the added costs of family occasions; birthdays, weddings, funerals and religious celebrations. For most purchases that a person makes,

23% of the cost of the item is taken by the state in value added tax. The working and middle classes are overburdened with taxes. This worsens the plight of the poor, the sick and the destitute. But it maintains or enhances the lifestyle of the super-rich.

A Property Tax, a Universal Social Charge and a Pension Levy were imposed on people with the collapse of the economy. Salaries and wages were reduced, extra working hours for less pay was introduced and tens of thousands of workers lost their jobs. Thousands emigrated. Most people's assets were reduced by more than 50%. Shares in manufacturing, construction, property development firms and financial institutions diminished to a fraction of their value. Many people were paying back loans to banks for property purchased during the boom years that was now valued at half the money paid to purchase it.

Hundreds of thousands of people were in serious negative equity. Young people seeking to buy a house could not get the money from banks that had collapsed or did not now have the money to lend. Those who wanted to sell their home couldn't do so because the money they'd receive would not pay back the loan so they were left paying a mortgage for a property that someone else possessed.

During the Celtic Tiger period people were lured into a fairyland, inflated economy; spending money they didn't have. They were given false and unrealistic promises. They were led like sheep to the slaughter by capitalist predators. This was not an accidental collapse of the economy any more than the unprecedented surge in the economy during the Celtic Tiger years was accidental.

The Irish government became impotent when the banking system collapsed. Outside help was sought. Three different international financial institutions came to take control of the economy, The European Central Bank, the International Monetary Fund and the European Commission. This three-some was known as the Troika. It was through their demands, in return for an €85 billion loan, that austerity measures were introduced.

Prior to this collapse we were hailed as the model country of Europe for our developing economy. We were the Celtic Tiger. What an image, a magnificent, powerful creature, and the people were led to believe that they were a part of that beast's power, vibrancy, self-reliance and strength. The image of a Tiger bounding effortlessly through the land was a most attractive, absorbing caption. Most people celebrated and took pride in being on the Tiger's back. The old Irish proverb portraying a person who is having a lot of success and luck is; *ar muin na muice which* translates to, *on the pig's back*. Notice that gigantic leap forward from the humble grunting pig to the all-powerful majestic tiger.

The phrase constantly used by our Taoiseach of that period and by many business people was: *going forward.* It was like going into battle; charge forward fellow-countrymen, no more looking back; we are on the Tiger's back, holding a sword of gold, leaping ahead. There were some people of course, the forgotten ones, who never even got a glimpse of the Tiger. Yes! We now know that this was no more than a mirage, a capitalist created economic illusory monster, a luring image to suck Ireland into Europe and the capitalist western world. Europe was moving from a war torn continent to a market driven, blow and burst (inflation and deflation), capitalist controlled western alliance with no need for boundaries. The power of money has become greater than the power of the sword; money and wealth surpasses the power of the most sophisticated war missiles

The people (the plebs as the capitalists would prefer to call them) were, not asked, but compelled to bail out those that caused the disaster. The people reluctantly accepted the imposed punishment. Austerity measures, penalising every man, woman and child in the state were introduced to pay back the €85 billion that was borrowed from the Troika to finance Government and pay Bank debts incurred during the Celtic Tiger years. After two years of austerity, for many people an unbearable two years of hardship, deprivation and poverty the Troika left the country. That super-controlling lending body was satisfied that the debt was being repaid; the plebs had been lured into accepting the blame and paying the price.

However, the exit of the Troika saw the imposed austerity measures remain; the Universal Social Charge, the Pension Levy and other measures introduced that inflicted hardship on those who were not responsible for the economic collapse. The ordinary working people continued to pay, many were forced to emigrate, many took their own lives and many were left homeless and destitute.

Not only did austerity continue, but a Property tax was introduced. Value Added Tax rates were increased. This caused all consumable items to rise in price. Because the people accepted all those austerity measures with very little resistance the government decided to introduce a water tax.

This was a step too far. Reckless governments including the main opposition parties in the Dáil surrendered our independence to Europe. Thirty pieces of silver and the chants from European capitalists to the Irish government leaders continued: *Crucify them, crucify them.*

Eventually, the people rebelled chanting: *No way! We won't pay.* The government then passed legislation to take the Property Tax from the people's salaries, wages, pensions etc.

This stealth-like draconian measure, compelling people who were reluctant to pay this tax, was provocative. It was hurtful and nauseating particularly since it was the property bubble created during the Celtic Tiger years by Government, banks and property speculators that caused the problem. The sovereignty of the people has vanished, their independence gone. They are treated as slaves as pawns in the wealthy, greedy capitalists' agenda.

There have been attempts by individual states to control and dominate Europe in the past. Frederick The Great of Prussia, Bismarck and Hitler of Germany, Louis 14th and Napoleon of France and in earlier times the Roman Emperors. Now it is a combined European capitalist collusion led by Germany. In the past all the attempts to control Europe were done through war, provocation, aggression, occupation and forced possession.

Today control and domination is done through diplomacy, persuasion and collusion. Monetary domination and wealth reserves are used as a weapon to achieve control instead of artillery. This operates through threats and bribes, supplies and sanctions; making available or withholding finances and resources. This is considered to be bullying in the work place but at the top level it is accepted as diplomacy.

Why the introduction of a water tax? The American war of Independence and The French Revolution were ignited when their governments adding taxes to tea and salt respectively. The people were already over-burdened with taxes. They rebelled. Revolutions changed the governance of both countries. The Irish people, despite a long history of protest and militancy, were now mollified by an overall improvement in living standards and a 'divide and conquer' approach that was well tested, established and operated in the western world.

The introduction of the water tax was aggravating; it unified many people and led to protests. In Ireland there has been no genuine attempt made to encourage people to conserve water. The water tax was seen as a revenue collecting exercise and a symbol of power and domination, not a genuine effort to conserve and improve the quality of our water. *Where will we get the money from?* This summarizes the government's attitude to water conservation.

CHAPTER 1

For Richer or for Poorer - All the Days of Your Life

According to a report on the cover page of the Irish Times on Thursday May 14 2015 the richest 250 people in Ireland are worth a combined €75.03 billion. They have seen their wealth increase by 15.9% in the last year. This newspaper was quoting figures published in The Sunday Times the previous week. The published figures of this rich list show that Ireland is currently home to 13 billionaires, who have a combined fortune of €37.89 billion, and that the net worth of the country's wealthy elite is now significantly ahead of that recorded at the end of the so-called Celtic Tiger era in 2008.

Denis O'Brien who is 57 years of age is believed to be worth €6 billion. He is the largest shareholder in Independent News & Media and is involved in enterprises such as the Topaz filling station chain, radio stations Today FM and News talk, and Siteserv, the purchase of which is at the centre of controversy and ongoing investigations. The two ahead of him on the richest list are Hillary and Alanna Weston owners of Brown Thomas

Department stores and Penneys nationwide clothing stores, and Pallonji Mistry, an Irish/Indian Construction tycoon.

David McWilliams, an economist, broadcaster, author and journalist researched and produced a television documentary called "Ireland's Great Wealth Divide" that was aired on RTE 1 on Monday evening September 21st 2015 giving us an insight on how the rich became richer while the poor became poorer during the period when our economy collapsed. That programme gave us a glimpse of how the wealthy live. They have private jet-planes, yachts, expensive cars, mansions with private swimming pools, Jacuzzis, snooker playing rooms, gymnasia etc. They usually have a number of properties at home and abroad. Their tax contribution to the state is relatively small and totally out of proportion to the contribution extracted from PAYE workers. That documentary showed how the earnings and possessions of the middle classes shrunk and the earnings and possessions of the working classes dwindled to close to poverty status.

We are informed regularly by government sources that Ireland has turned the corner. The country is back moving again. Yet, there are serious problems in the Health system with thousands on waiting lists for admission to hospitals for necessary surgery, hundreds of seriously ill patients on trolleys on hospital corridors awaiting admission to a bed in a ward, large numbers of children in classrooms in our cities and towns, a shortage of SNA services for children in need, a huge shortage of housing, daily evictions of people unable to repay mortgage loans to banks or rent to landlords and poor employment prospects for our youth.

When the Taoiseach, Tánaiste and minister for Finance are questioned about these failings of government they reply with another question: *where will the money come from?* Have they ever considered the possibility of getting some more money from the super-rich, those who are worth €75.03 billion and make it available to help the Irish people? They could encourage and stimulate home industries, develop our fisheries, forestry, farming and tourism.

There are homeless people dying younger than they should from lack of food and shelter. There are people with serious drug addiction issues and there is no help for them. There are people dying needlessly because they cannot get the medical assistance to keep them alive.

In The Sunday Business Post February 2nd 2014 Dr. Anthony O Connor writes that "lives are in the process of being lost, because people with curable conditions are waiting for months to be initially seen in an out-patients clinic, and for further months to access diagnostic procedures such as scans and endoscopies." In the same article he writes about Ireland is a country: "with the highest number of medical schools per capita in Europe but cannot adequately staff its hospitals without pilfering doctors from the developing world." "An ideologically right wing minister presided over the establishment of the most bloated and inefficient public body, the Health Service Executive (HSE) in the history of the state, where nine organisations were melted into one without a single demotion, redeployment or redundancy". Succeeding ministers for health, instead of administrating the last rites to this inept, soul-less body, have continued with wasteful, expensive resuscitation.

The Journal.ie gives some startling figures relating to our health system in issues dated September 24th 2014 and April 10th 2015 in particular.

The waiting list for hospital treatments has swelled by almost a third since February the previous year. According to the figures in The Journal.ie on 10th April 2015,

25,000 children are waiting for their first outpatient appointment and 10,000 of them are waiting for more than 6 months. Nursing home waiting lists are spiralling out of control and there is a twelve-month waiting list for MRI scans putting lives at risk (October 2014).

The Journal.ie of March 31st 2014 says that 15,776 children are waiting for speech and language assessments and almost 3000 children are waiting more than a year for speech and language therapy. More than 3000 young people are waiting for mental health referrals (July 7th 2014). Thousands are going blind while waiting for cataract surgery. Again, quoting figures from the Journal.ie, figures obtained from the "Irish-health clinic":

"5 people go blind in Ireland every week. Over 220,000 people in Ireland are blind or visually impaired and this figure is expected to jump by 20% by 2020, the National Vision Coalition has warned. It is calling for the immediate implementation of a national vision strategy in an attempt to reduce this figure and save the healthcare system millions of euro.

Figures show that in 2010, blindness and vision impairment cost the State around €205 million, however proper investment in this area could save up to €76 million every year."

A reporter with The Irish Daily Mail, Leah McDonald, wrote on Monday 29th June 2015 that a top surgeon and consultant in transplant surgery, Dr. David Hickey, told a crowd in Liberty Hall that: *"we need to get active and agitate to protect our essential services"*. He said this in light of the closing down of the pancreas transplant programme at Dublin's Beaumont Hospital.

The title of an article by Petrina Vousden on page 8 of The Irish Daily Mail on Wednesday May 14th 2015 reads:

"EIGHT PAINKILLERS A DAY WHILE SHE WAITS FOR OPERATION.

A teenage girl from Raheny is on 8 painkiller tablets a day plus anti-inflammatory drugs, as she waits for life-changing surgery". This young girl suffers from scoliosis a crippling, progressive curvature of the spine. The waiting list for surgery at our Lady's hospital for sick children has increased this year. "This girl's curve is now at 70 degrees. Psychologically and mentally, it affects her. She hates herself, she is depressed".

Another article in the same newspaper on the same date written by Fiona Looney states "Over the past few weeks, mothers presenting their new born babies for the BCG - one of the blocks of the childhood immunisation programme and a drug that has effectively protected Irish children from TB since the start of the 1950s – have been told that there is no supply of the vaccine currently available in this country."

In the previous day's Irish Daily Mail Lea Mc Donald reported that: "Eight patients suffering with chronic pain took their lives in the past two months as they could no longer cope with the 'devastating' condition, a support group has revealed."

Ferghal Blaney political correspondent with The Irish Daily Mail reports on the 4th November 2015 that a couple in their 90s were left languishing in A&E on hospital trolleys which sparked a furious letter from a consultant. *"The husband, who suffers from Parkinson's, had been on a trolley at Tallaght hospital Dublin for 29 hours and his sick wife for 7 hours"*. The outraged consultant said the couple had been left 'to fester' in inhumane and torturous conditions. The consultant warned of an imminent death as a result of overcrowding. Patients on trolleys have 'no privacy, or dignity' they are subjected to 'constant noise torture, light torture and sleep deprivation which are a clear violation of human rights. Furthermore, he said the situation has not improved since it was reported in June how a 101-year-old had spent 26 hours on a trolley at the same hospital.

Some of the damning extracts from the consultant's letter are quoted by Ferghal Blaney: -This is *"an example of how a dysfunctional system disgracefully treats some of our senior citizens, some of the most vulnerable in our society, being allowed to fester on a trolley initially in a cubicle then transferred to a non-designated patient area conduit in the Emergency Department."* A second extract is condemnatory: *"this is most unacceptable and 'a slap in the face' for our most weak and vulnerable an even bigger 'slap' to those who stated this would never happen again"*

The Taoiseach's reply to this was that it is unacceptable. Both the Taoiseach and Minister for Health have been following the same line of reply to questions relating to problems occurring under their watch: agreeing with the unacceptability and sheer disgracefulness of such occurrences. They gloss over their responsibility in those unacceptable situations. They speak of their shock and abhorrence with the system as if they had no responsibility whatsoever in those areas. They even go as far as to ask: *who is responsible for this?*

In The Irish Times newspaper on Wednesday January 6th 2016 we read: *"Hospital waiting lists rise again to record level."* Elaborating on that, the report continues: *"The number of people waiting over 12 months for outpatient appointments increased to 85,130 from 83,347 in April 2015 despite Minister for Health Leo Varadkar's pledge that tackling long waits is a key priority for the government".* The Irish Times informs us that *"Tallaght hospital has over 10,000 patients waiting on outpatient waiting lists for over 12 months and Galway University hospital has almost 9,000"*

In The Irish Daily Mail on the same day January 6th 2016 Leah McDonald and David Raleigh report on how "patients spoke of the bedlam they encountered in one A&E yesterday as the numbers waiting on trolleys and wards nationwide soared to almost 560. A prominent emergency consultant hit out at the over-crowding problem, calling it a scandal that amounts to gross governance failure on a massive scale." The article continues: *"The daughter of an 81 year old lady, who last night had been waiting at the hospital since midnight on Monday, told of the bedlam they saw –*

including the shocking moment a makeshift coffin carrying a deceased patient was wheeled past them..."

With such problems in our health system can we say that Ireland has turned the corner?

People are being evicted on a weekly basis because they cannot afford to pay their mortgage or rent. Children with disabilities are not receiving the education they require because of lack of funding to schools and reduced staff levels, teachers and special needs assistants.

With the health system in chaos, a homeless crisis in our major cities and the suicide rates reaching unprecedented levels the Government became engaged in two referenda that were less pressing than rectifying the health system, poverty, homelessness and the plight of children with special needs. Those two referenda acted as diversionary tactics to focus attention away from inefficient, ineffective and simply bad government.

There are hundreds taking their own lives in this country every year. The suicide rate among young Irish girls is the highest in Europe and among young males it is second highest. In both young men and young women the suicide rate is more than twice the European average. Depression, due to poor future living prospects, obviously plays a large part in the aforementioned unacceptable statistics. This is happening in a country that is bubbling with wealth and rich in natural human resources. In the Sunday Independent June 21st 2015 the title of an article by Ronald Quinlan reads: Male suicide rate jumped by 57% during recession the subtitle reads; new study concludes there were 561 more self-inflicted deaths in five years of the crash". The article states:

"It's been said that recessions hurt, but austerity kills.

The findings of research by academics at UCC into the effects of Ireland's economic crash would appear to bear that out.

According to the study - the results of which have just been published - suicide rates for Irish men jumped by a staggering 57% from the onset of the crisis in 2008 to 2012.

All told the research carried out by Dr. Paul Corcoran and his colleagues found that in the five years of the recession there were 561 more deaths by suicide than there would have been if pre-recession trends had continued."

This is a terrible indictment on Irish politicians where greedy capitalists, bankers and speculators were encouraged and allowed to flourish at the expense of peoples' lives. This damning legacy of greed and neglect has a resonance and harrowing similarity with the great hunger of 1845 - 1848 in Ireland. During that period hundreds of thousands emigrated, thousands were evicted from their homes and land, they had not the means or strength to emigrate and hundreds of thousands died of starvation. It is sad when power and money take precedence over, not only basic living requirements but even over people's lives. We know that in the underworld of crime and gangsters a person's life means little by comparison to money and loot, but governments are established to protect people and improve their standard of living.

Ireland is a rich country. It has lots of quality land and abundance of fresh water. It is surrounded by seas that are laden with fish. We have beautiful scenery, we have wind and ocean wave energy and we have a small population to feed and care for. This is a country where the rich continue to get richer and poor become poorer. The super-rich, by-and-large, spend the cold winter months abroad in sunnier lands. They come back in the summer months, like the cuckoo, to lay their eggs in other bird's nests. Then they fly off and leave the home bird to do all the work. This gives them a better life style and they can through the luxury of living abroad avoid paying their fair share of taxes in Ireland. Taxes are taken from the poor. If they can scrape enough Euros up they can emigrate. If they cannot do that they are doomed by this unequal, unfair system where threats are frequent, the latest threat being: *if you don't pay we will take the money from your income be that from employment, pension or social welfare, or, we will take it out of your bank account.* Rob the poor to pay the rich. Phil Hogan made the infamous threat saying if people do not pay the water levy their water supply will *"be reduced to a trickle"*

AUSTERITY THE REAL ISSUE

Comments that some people expressed in social media:

- *I have had and still have personal reasons to have to visit hospital on a daily basis and the issue is not infra-structure, the facilities are there, the beds are there but the nurses and doctors are not there.*

- *Recently, on the 5th floor of the Mater Hospital half the beds on that floor were closed, there were 3 staff members looking after 22 patients overnight, one was an experienced nurse, one was a recent entrant care assistant and the third was a brand new inexperienced care assistant, the two care assistants were agency staff and had poor English language skills. They had great motivation I cannot fault their intentions or work rate but downstairs in the A&E department it was full of people on trolleys waiting to be admitted to a bed.*

- *The stock of this Country, the people are expendable. Why? Because they are easily replaced, the source of Government funding and by extension Bank funding is the stock, tax-paying individuals. It is a never-ending source of funding. They can be squeezed and dispossessed at the whim of whatever vested interest has the power.*

- *Does our FG/Lab government really care about Homelessness or Social Housing?*

- *Since a young man died 50 yards from Government Buildings and the several days of fury following that terrible event, do you hear any government official commenting on Homelessness or its causes?*

- *Eviction is becoming the main cause of Homelessness and it's not the traditional profile of a single young person becoming homeless, its entire families, young children whose formative years are being completely disrupted, moving from one temporary accommodation to another, disrupted school attendance, degenerating diets due to a lack of cooking facilities in temporary accommodation and the increase in family unit breakdown caused by stress.*

- *The longer all the conditions above exist, the more they become the norm and accepted by society but they are all totally unacceptable.*

- *EVICTION comes in many forms. In 2015 Ireland, it is no longer simply the traditional arrival of Sheriff's and Bailiff's although this does still happen in far too many cases, but also there is the psychological Eviction, were a Bank will engage in a personal campaign of harassment by phone, by letters, by threats of Court and actual Court proceedings until a point where the battered family simply gives up and either surrenders or abandons the Family Home, it is war of attrition which is condoned by Government, the Courts and the Legal Profession and it is largely ignored by the media.*

- *Ironically, the above mentioned segments of the professions are the same groups who trot out the "for the greater good" mantra and also the same professional groups who harangued and cheer led the property bubble which led to financial crisis in Ireland and around the world and finally, the same groups who make vast amounts of money on the way up in a boom and again on the way down in a recession*

When we got our independence we thought that we would manage our affairs and treat our people with respect and dignity. Our constitution had clauses protecting families, giving citizens of the state a sense of belonging, a patriotism and loyalty to their country. A buy Irish campaign gave a sense of loyalty to our workers and our indigenous industries. We painted our telephone booths, our letterboxes, public gates to heritage sites green to distinguish us from our colonial past. Aer Lingus had green airplanes and the shamrock as a logo and even our buses and trains were painted green. A pride in our country was established. We discarded the red of the British Empire.

Now blue is slowly becoming the colour in line with Europe. We have again lost our identity subsumed and submerged once more into a new empire. We have lost the vision of self-sufficiency that was envisaged with the establishment of the Free State. Of course we must embrace pluralism. But we should not get lost in the pack. We must preserve what is best about our country.

The ordinary people of Ireland have suffered for the past six to seven years because of austerity imposed on them. In a programme on TV3 called *The People's Debate* hosted by Vincent Browne, Martin Heydon a Fine Gael TD from Kildare in summing up his contribution to the debate said, that the people of Ireland made sacrifices to get the country back working again. Vincent Browne's instant, spontaneous and condemning reply was - No! - should it not be:

Your government sacrificed the people. The people didn't make sacrifices; they were sacrificed.

Vincent Browne in his late night debate programmes has consistently, vociferously and forcefully defended the voiceless people who have been the victims of the recession. He, and other journalists, commentators and broadcasters have spoken on behalf of voiceless, downtrodden people. They, and not government, have represented the people during the harsh times of the recession.

CHAPTER 2

Uisce Faoi Thalamh

The literal meaning of the Irish phrase *uisce faoi thalamh* is; under-ground water, but it can also mean secretive or under-handed.

The new water meters that have been installed in many parts of Ireland have the word WATER written in the centre of the cover, with the word UISCE given a less prominent position below it. UISCE is the Irish for water. UISCE BEATHA is the Irish for the water of life. Water is a life sustaining necessity. All creatures, by virtue of being born on earth, have a birthright to this natural resource.

Humans have very little interest in exploring planets that they believe do not have water. The first thing space explorers seek on the moon or on a planet is water. Where there is water there is life and where there is life there is water. Ireland has so much water, lakes, rivers, seas and rainwater that we could say; Ireland and water are synonymous. When a foreigner is asked what he or she knows about Ireland the reply usually is; *it is a green country with lots of rain.*

The first public water regulator stop-cocks or valves outside each house had cast iron covers with a lovely decorative Celtic design and just the word UISCE on it. This was a symbol of our identity. It linked us with our Celtic origins. It had a permanency being made of iron. It was designed and made in Ireland and had the Irish word UISCE embossed on it.

Then came two less elaborate designs but they still had only the word UISCE embossed or inscribed. They too were made of cast iron; one was circular in shape the other square. The water system was managed by the local county council and financed from general taxation.

As an essential life sustaining substance this was considered such a necessity that entitlement was a person's right and so had to be provided by the state. Recent covers have the word water inscribed in the centre of a circular, uninspiring black polyvinyl chloride (PVC) lid.

Some have the word UISCE omitted entirely. This symbolizes our loss of identity. Like the spire in O' Connell Street in the centre of our capital city this further signalled and symbolised, not only a diminution in our Irish and Celtic identity but also, a failure by government to seek, develop, foster and appreciate Irish designs, designers and craft-workers. It was also a missed opportunity to establish Irish jobs and give employment to the many unemployed in our country.

The meters are inserted in a hole on the pavement or on the road outside the house being monitored. They are about 30cm down and the hole is narrow about 15cm wide at the cover. It is difficult to remove the tightly fitted black PVC cover. It can only be removed with the aid of a sharp tool and that instrument is not supplied by Irish water (Uisce faoi thalamh) Furthermore, a person cannot read the meter without stooping down to almost ground level, thus discriminating against old, invalided or physically handicapped people. They cannot be read after dark without the aid of a lamp. They are not user friendly. They can be read externally by those billing the customer but the customer is not afforded that privilege to read his/her own water usage. No smart reading device is made available to the consumer.

Surveillance devices used to monitor a citizen may be in breach of the person's right to privacy and contravene the spirit, if not the terms, of data protection legislation. In the case of electricity meters and gas meters they are attached to the property and the property owner gives permission to the supplier to read the meter. Not only are water meters placed outside the proprietor's property but are placed underground and in very inaccessible places for the customer to check as pointed out above. This is incompatible with the philosophy of openness and transparency in the supply and delivery of a necessary public service. In many cases five or six recording water meters are clustered together on the footpath making it difficult for each house-owner to identify his/her particular recording meter. This could be construed as uisce faoi thalamh.

WATER METERS

Stooping to peer down the hole for a reading
Head in the hole, arse in the air
Bending still further arse rises higher
Panting and blowing such stress on the breathing

Now! He bends down peeping with lamp
Belly hits pavement, glasses fall in
Cursing and swearing now every where's dark
Reaches for glasses; Oh shit! Leg's got a cramp

Next she stoops down taking the lamp
She bends even lower to have a good look
Her bra string snaps and her boobs have a peep
Cars passing splash water then halt at the ramp

All down the road they're doing the same
Everyone swearing and nobody paying
The bills are a coming but the people refuse
To concede or surrender: Oh! What a shame.

Yea! Come April Fool's day, who's going to care?
Taoiseach and Tánaiste arsing about
His head in the hole: Her arse in the air
As they curse big Phil Hogan living out there.

Water is our greatest natural resource

As noted earlier when astronomers, scientists and space explorers set out to reach and explore a planet the first thing they seek is water. If there is water on a planet then there is life. Oxygen is one of the two elements necessary to produce water. This is also the essential element to sustain life. One fifth of the earth's atmosphere is composed of oxygen. Without the oxygen in the air all earthly creatures would die. Without the oxygen in the water all Aquatic life would die. Water would not exist without oxygen. All vegetation needs oxygen which is taken in through the foliage from the air above and through their roots from the water below. Oxygen is a God given life-sustaining element. For the atheist it is an evolved element that aids the development and sustenance of life on planet earth. Theist and atheist agree and unite in their acceptance of the fact that water and its essential elementary component oxygen are natural resources that we require and are entitled to as creatures of planet earth. They are given to us free from nature or from God. It is a human responsibility to care for our water and make it available to our fellow humans. What some humans have done to damage that life sustaining essential resource is shameful and appalling. To deprive or place a charge for water on our fellow humans, our young, our old, our sick and infirm is unacceptable. All creatures of the earth are entitled to this necessity.

In Iceland the people benefit from their two great natural resources geo-thermal energy and aluminium. 90% of domestic homes, apartments and retail outlets are heated by means of this natural resource. It is also a great tourist attraction there. The Blue Lagoon and the hot springs, amid extreme cold weather conditions, attract hundreds of thousands of tourists every year. Iceland is one of the top producers of aluminium in the world. In Norway natural gas, oil and snow are their natural resources. They exploit those natural resources to bolster their economy. The French take great pride and use to the full their natural resources, wine from their luscious sun-soaked grapes and fine silk from the industrious silkworm that weaves silk for them on their Mulberry trees.

Ireland's greatest natural resource water is available in abundance. It is given to us in voluminous quantities. We are an island surrounded by water. We have thousands of fresh water lakes, rivers and streams. We have numerous natural springs. Politicians at home and abroad want to charge the people for this natural resource. In very hot countries of the earth where very little rain falls, water is scarce and difficult to obtain. They may have other natural resources in those countries that people in Ireland do not have.

Many uncaring and greedy humans and government leaders all over the world have polluted countless rivers, lakes and oceans of the earth with chemicals and waste products. No other creature on earth has done such damage to this natural life-sustaining essential resource. Industrialists maintain that in the name of progress they are helping the human race. But in reality it is greed and self-interest that they pursue. The damage they perpetrate is treated as accidental, collateral damage. Nuclear plants, Smokey fossil fuel burners, incinerators for waste disposal, gun and bomb making factories, tobacco Industries, plastic manufacturing factories, fluoride added to water have not been to our benefit but to our detriment and we are expected to pay additional taxes for such destruction. They call it development.

Employment creation should not be an end in itself. Governments want to have full employment to keep people occupied. Their reasoning is to keep people from being a threat to the stability and cohesion of the state. Governments also receive a huge dividend in tax income from those employed. Not only do they take income tax from workers but also an enormous amount of indirect taxes on goods that each person purchases. This involves workers being forced to pay back most of what they earn.

People should work to make life better for themselves, their children and society. People need to work for self-satisfaction for social reasons and to contribute to society. Those should be the goals for working. Work should not be used as a means of controlling people, of using them for scientific, industrial, experimental or financial gains for domineering, controlling bureaucrats, financial institutions or greedy governments.

We, in Ireland, could sell water to all of the wealthy countries that have an insufficient supply of water to serve their people and give it free to some underdeveloped countries as part of our responsibility to help our fellow humans.

As the need for water increases on earth Ireland is well equipped to cater for this requirement. We will need to exploit the advantage of having a wealth of water to encourage industries that are highly dependent on water. In 1983 Ireland through our minister for trade John Bruton offered water to Saudi Arabia for $5 a barrel. However, at the time his proposal was rejected mainly because Saudi Arabia did not want to become dependent on us, or, on any nation for its water needs.

Human beings in a civilized society are also entitled to water, to a basic primary education, an adequate, satisfactory health system, a protective impartial policing service, a social welfare system and a civil service to manage essential government departments.

Water is a necessity for living. The average human body is composed of 60% to 65% water. We need a daily supply of water in order to remain healthy. We need water to cook our food, to wash ourselves, our body, hair and teeth. We need water to wash our clothes and our cooking utensils. All creatures on earth need water; fish, animals, birds and insects need water. Those creatures do not need expensive clothing, footwear, jewellery or processed food. They do not need to go on exotic holidays. They do not need information technology either i-phones laptops or i-pads. They don't need tobacco, alcohol or other luxuries of life that many humans enjoy. But they need air and water; *that is their birthright. It is our birthright too.*

CHAPTER 3

Irish Water – The Quango

We don't need a quango (**quasi-autonomous non-governmental organisation**) called Irish Water. Each county council should have the expertise to deal with water issues, including leaks from the system, purification and sewage treatment. Knowledge of pipe structure, location and composition of piping ought to be logged and handed on from generation to generation of workers down through the years. Paying consultants 86 million euro was an unnecessary expense. Those workers already employed by county councils should have been given the extra allowances and heavy-duty equipment or given the finances to hire equipment to carry out all necessary repairs and improve the system. The cost of setting up Irish Water to placate European politicians and relieve the Irish government of the responsibility of dealing with water supply and conservation was enormous. This cost is an unnecessary extra burden on the Irish people who worked hard to build a country for themselves and their children.

Ben Haugh writing in The Irish Mail on Sunday January 10th 2016 presents the lead story on the cover page with the heading: Irish Water's €20k a day on Consultants. He reports as follows: "Irish water spent almost €20,000 per day on consultants in the first eight months of 2015 – and that's on top of the €86 million it initially budgeted for external help to set up the company".

There was controversy two years ago when Irish Water's chief executive officer (CEO) John Tierney let slip that €86 million of the €172.8 million setup costs for the utility was being spent on consultants.

Crucially, in 2014 Mr. Tierney insisted those costs would fall once Irish Water began 'hiring in' expertise. But the company is continuing to shell out millions of euro on consultants. The company calls these fees 'ongoing costs' and insists they are different to their establishment costs."

The company has two distinct headings for costs 'establishment' and 'ongoing' writes Ben Haugh and he continues: "... documents released under FOI (Freedom of Information Act) to the MoS (Mail on Sunday) show that A&L Goodbody was paid about €25,000 in July 2014 for 'legal services' as part of the establishment of Irish

Water. A&L Goodbody also got another 11 payments totalling €60,000 in the same month for legal services – listed under 'ongoing costs'.

In December 2014 Accenture received about €660,000 as part of the setup IT support costs. In the same month it also received €45,000 in two payments listed under 'ongoing costs' for function support.

In November 2014 Irish Water made five payments to Ernest & Young as part of its setup totalling almost €25,000. The services are listed as functional support, programme support and systems development and support. In the same month it made three payments in respect of services, such as procurement optimisation programme, totalling about €48,000 to the firm under 'ongoing costs' ".

How can these costs be justified?

Workers are being blamed for the damaged economy. They are asked to pay extra taxes to undo the harm caused by the rich and greedy people who do not pay their fair share of taxes here. The ordinary citizen is brought to court and charged for tax evasion. There is one law for the rich and another for the rest of the Irish citizens. A company controlled by one of those rich guys is given charge of installing water meters.

How was Irish water established as a utility company?
A construction company called Siteserv became bankrupt. It owed €150 million to Anglo Irish bank. That bank itself was bankrupt and was taken over by the state. Siteserv was a profit making company but had borrowed more than it could repay. The company had to be sold in order to recoup some of the money it owed. It was sold to Millington, a Denis O'Brien controlled firm by the Irish Bank Resolution Corporation (IBRC). The IBRC was formerly the Anglo Irish Bank. Independent TD Catherine Murphy, through the Freedom of Information act, sought details of the sale. Every effort was made to block her probing and bringing to the public how this company was sold by IBRC and bought by Millington. Catherine Murphy TD highlighted the results of her findings. €100 million of the company's debt was written off. This €100 million became the Irish people's debt as the state now owned IBRC. €5 million was distributed among the shareholders of the company. Denis O'Brien paid the remaining €45 million for the company.

This was the company that became involved in the installation of water meters. But before work on this new water company began €86 million was spent on consultant's fees for their advice on the establishment and development of Irish Water. So the Irish people are faced with a starting bill of €186 million. The residential property tax was introduced to pay for local services but now the most important of local services is not being covered by this tax, they say, and a new water tax is introduced. Groups opposing all the austerity measures came together to protest, the Austerity Alliance, Right 2 Water, People Before Profit, the Socialist Party and Sinn Fein. They organized a protest march through O'Connell Street where over a 100,000 people took part on the 11th of October 2014. On November the 1st demonstrations throughout the country brought over 150,000 to the streets and another demonstration in Dublin on the 10th of December a mid-week protest brought about 60,000 to the gates of the Parliament House 'An Dáil'.

Brendan Ogle leader of Right2Water and Unite Trade Union coordinated the work of anti-austerity activists, voluntary workers and historians in getting the message bellowed out on public platforms, facilitating solidarity and uniting people in their fight against water levies and otherpenalizing austerity measures.

Richard Boyd Barrett, leader of People Before Profit, and Joan Collins have worked tirelessly to highlight corruption in politics, oppose the establishment of quangos and help to bring to the fore the plight of working people. Boyd Barrett, along with Sinn Fein councillor Shane O'Brien organised protests in Dun Laoghaire against the introduction of water rates.

The Socialist TDs Joe Higgins, Clare Daly, Paul Murphy and Ruth Coppinger were main leaders in the protests against austerity and water taxes. They prepared motions and spoke vociferously in the Dáil on how the rich were living off the taxes paid by the working man and woman.

Sinn Fein TDs particularly, Mary Lou McDonald, pleaded passionately for the abolition of property taxes and the scrapping of water rates. The vast majority of independent TDs were opposed to the introduction of water rates and called for the dismantling of this new quango *Irish Water*. The main opposition party Fianna Fail stood idly by just as Fine Gael and Labour did when Fianna Fail and the Green Party were in government. They watched the corruption and witnessed the collapse with little or no retort. Fianna Fail eventually joined in and called for the abolition of Irish Water and this unjust tax.

Governments must listen to the people. <u>Why tax water?</u> The Government's reply is with another question. The people, in a democracy, ought to ask their government leaders' questions and not the other way round. The Government's question to the people is: *Where will we get the money from?* And, who is going to pay for the supply and maintenance of water and the disposal of waste? The people reply with a simple and straightforward answer we have already paid in general taxation contributions. Services provided through general taxation should include this service as it always has done.

29 "Irish water" staff members earn over €100,000 each and as the Journal.ie informed us 27 senior managers are given a car allowance too and that nearly 100 staff could earn 15 per cent of their salaries in pay-related bonuses. Director John Tierney

The senior staff in Irish Water will be paid over €3 million a year.

The figures reveal that there is a staff total of 310 at Irish Water. 127 of the staff earn over €70,000 each year; 42 earn between €70,000 and €80,000; 35 earn between €80,000 and €90,000; 21 between €90,000 and €100,000; 19 between €100,000 and €125,000; 9

between €125 and €150,000 and the Managing Director John Tierney earns over €150,000

94 employees of Irish water are eligible for either 14% or 15% performance related bonuses on top of their salary; 165 are entitled to 6.5% performance related award pay and 50 are given an award of 2.75% while there is a pay freeze on public service pay'

Minister Phil Hogan was asked for details of staff eligible for car allowances, with the answer revealing that 27 senior managers were given €10,500 a year for using their cars on the job.

Nine senior executives receive health insurance with VHI, at a cost, according to a quote from the VHI website, of around €2,631 a year each, totalling €23,679.

There was a performance-related pay system introduced for workers in Irish Water

Speaking at Leaders' Questions in the Dáil, Mícheál Martin raised the issue of bonuses being paid to staff at Irish Water, the new State utility that has been the subject of controversy since it was set up. He asked Mr. Kenny why the government approved the bonuses.

The Taoiseach described the payments as "performance pay" and said that they would be paid on the basis of staff in Irish Water "achieving a particular set of criteria and targets".

He said he has asked the Minister for the Environment "in the interests of transparency and accountability" to have the Irish Water chief executive officer, John Tierney, supply him with the criteria for the payments so that "everybody will understand what is involved if an employee is to achieve an output of work that would allow him or her to achieve a performance pay rating".

He said that John Tierney will be happy to lay out the information and pointed out that Irish Water "will be scrutinised by the Oireachtas here and by civilian society because it's subject to the Freedom of Information Act". But Mícheál Martin responded saying: "The bonus culture is back, is basically what you're saying, and, it's alive and well in Irish Water."

A series of Tweets followed. I will give a brief summary of some of those tweets.

- *Mr Kenny & Co took the Christmas bonus from the elderly and needy. What with the austerity measures his government put in place over the last five years or so, I believe firmly that my 85 year old mum has performed more than admirably in meeting her set 'targets', yet her 200 euro bonus was denied. So she came to stay with us over Christmas and that saved a bit on the oil.*

- *Irish water management do not under any circumstances deserve a penny more…but it's not a 'bonus', it's 'performance pay' – just like it's not a 'tax', it's a 'social charge'… weasel words…*

- *The Taoiseach as usual doesn't answer the question and it's correct the bonus culture never went away. Ask the government ministers' special advisors who are paid above the government's own salary guidelines. Ireland is getting more Orwellian day by day to different strands in society with different rights and entitlements.*

- *I am truly sick and tired of official Ireland wasting my money. Private companies pay bonuses when the company is (A) making money and (B) doing well enough to maintain investment and pay a thank you to the staff. Irish Water has no income and so far has made a loss of 180 million Euros. Even allowing for the fact that public sector workers should not be getting bonuses in the first place, why would Irish Water (a loss making monopoly) be paying bonuses on top of ridiculously over the top salaries?*

- *It's time for the Irish people to rise up and reboot our democracy.*

- *Hypocritical in the extreme by Mícheál Martin but he is however correct. Even a stopped clock is right sometimes. Almost half way through this coalition's term and little or nothing has been done to regulate bonuses, top-ups, expenses, unessential quangos, enforcing tenders are followed up correctly, upward only rents, overspending by councils and it's only going to get worse now that the steady hand of the troika is gone. It's time to buy an election.*

The leader of Fianna Fáil said that bonuses had been approved despite the objections of government ministers and he claimed that the coalition went against its own advice in setting up Irish Water as a subsidiary of Bord

Gais, something a report from PricewaterhouseCoopers had advised against.

The Taoiseach said that the present situation with water provision cannot be allowed to continue, saying that 18,000 people on public water supplies have to either boil their water or have restrictions in place and that supply in Dublin is at 96 per cent "This cannot continue and it shouldn't continue", he said.

€86 million paid to consultants to advise on issues relating to Irish Water is dead money. The government feeling under pressure from mass protests against Irish Water decided to give €100 euro bribe to those who signed up to pay Irish Water, they called it a conservation grant but you can use as much water as you wish and waste water if you wish because you will still pay the same amount. That's hardly conservation and the same grant is given to the person in a mansion who has a swimming pool, Jacuzzi, hot tub, garden water sprinklers etc. as to the single person living in a humble dwelling with a water harvesting system

Those who Boycotted Irish Water and refused to pay this unjust and unfair tax were doing what people in County Mayo (the Taoiseach's county) did to Captain Boycott in the 1870s. The people of rural Ireland, in the second half of the nineteenth century, after The Great Hunger from 1848 were treated little better than slaves; many were evicted from their homes and lands. Some were left in their cottages but had to pay a rent to the landlord while their lands were taken off them. They were then forced of necessity to work on landlords' farms in order to feed their families and pay the rent.

Tenants and workers on Captain Boycott's estate refused to work for him; they ostracised him. They refused to acknowledge or speak to him. Eventually, this type of ostracising of a person came to be known as Boycotting. This is how the word Boycott entered the English dictionary. People who tried to halt the installation of water meters saw Phil Hogan in the same way as the people of Mayo in the 1870s saw Captain Boycott but they were taking their protest a step further. This was seen as Hooliganism rather than Hoganism by the authorities. A number of protesters were taken to court and ordered by the court judge not to go within 20 metres of meter installation work. They disobeyed that order and subsequently five people were sent to jail. They questioned the authority of the judge. They maintained that the sovereignty of the people must be given some consideration.

CHAPTER 4

Water Conservation

We can do much more to conserve water in Ireland and at the same time ensure that every citizen receives an adequate supply of clean water. Why are we supplied with treated, fluoridised water to flush toilets, wash floors, cars and water gardens? Why are there still so many large cisterns flushing toilets? Successive governments have done nothing to conserve or look at ways to use water more efficiently. Whatever has been done to conserve water has been done by entrepreneurs or forward-looking business people and they, by and large, follow what is happening elsewhere. This leaves us trailing behind most progressive countries. Why not use rainwater more since we have so much of it in Ireland. Why don't we recycle water? Why don't we use waterless loos?

Tanks can be mounted on the roof of a garage or shed or installed underground to collect rain water, this would not diminish a premise's space and tanks can be of any size, Underground tanks require an electric pump to pump the water into an attic tank for distribution to toilets and for most of our domestic uses. They also need to have ultra violet filters fitted to kill bacteria. This system would be relatively inexpensive to operate and maintain when installed. Every property should have two water storage tanks. This requirement should be part of planning permission for all new dwellings.

Storage tanks should be large enough to last a family of four for two weeks or even longer with careful use even when no rain falls during that period. We should also use recycled water from baths, showers, kitchens, washing machines and dishwashers. This water should be stored in a separate tank to the rainwater. The water in that tank should be used to flush toilets, water the garden, wash the car and hose paths. It is also possible to have the rainwater harvesting storage tank installed with an automatic refill system from a main water supply when the water runs low in the rain storage tank. The government supply scheme should include a direct supply to a drinking water tap in every house. Grants should be made available to construct or install water-harvesting tanks. Grants should also be available to people in isolated rural dwellings to install bio-treatment sewage systems and to drill for quality spring water or to supply drinking water from a group scheme.

Drinking water should be filtered and supplied to all houses. Nobody in Ireland should be denied this essential necessity to healthy living. Governments ought to fulfil their obligation to their people. The people in a democracy are sovereign, Elected representatives must ensure that this obligation is adhered to and the cost must come from general taxation. Education on water usage and conservation should be included in all schools' curricula, starting at preschool level. This should be part of the Social, Personal & Health Education (SPHE) programme and complemented by modules in the Social, Environmental and Scientific Education (SESE) programme.

MANAGING WATER

The supply of water on earth will not deplete. It is in a cyclical, continuous ever-renewing process. It evaporates, collects as clouds in the sky and falls as rain. It then soaks into the earth thus supplying all vegetation including our large rain forests with water. The vegetation that is nourished with this water, supplies all creatures on earth with fresh air and food to sustain them. It cools the hottest parts of our planet and heats the coldest parts. Clouds protect the earth, humans and all creatures on earth from over exposure to sun radiation.

We don't need to ration water. We need to know how to manage it. Education on ways to conserve water with encouragement to manage water with incentives is much more effective than enforcement of laws which treat people as irresponsible, careless and wasteful. Two thirds of the surface of the earth is water. If we include the water contained in most objects and in all living creatures on earth as well as in vegetation, in bogs and in the soil, the earth is composed of over 90% water. When we consider that most of that water operates in a self-regulatory circulation system and that humans in positions of power are at fault for the damage caused to our water we must awake, speak out, protest and object.

People in Ireland should not be made to feel guilty about using water to keep their plants alive in hot summer weather or to wash the car. When the water is used prudently and for the purpose of keeping plants alive and healthy and to keep our environment clean it is not only acceptable but laudable and should be encouraged. Inculcating a guilt feeling in people is unhealthy and deplorable. Guilt feelings cause a lot of problems for many people. Education, information and incentives to improve people's behaviour ought to be the way forward. Blaming people for using too much water to flush toilets is incredible when duel flush or air suction type toilets should be a necessity in building all dwellings, including offices and public facilities.

Of course water must be treated; filtered and in some cases purified to ensure that it is safe for drinking. Sewage too, must be taken away from towns, cities and dwellings and treated. Human toilet waste should not enter our streams, rivers, lakes or seas. It costs money for quality filtration systems, purification and sewage treatment. The question repeatedly asked is *where will the money come from?* One can ask such questions about our primary schools, public hospitals and other public services. The answer is as it always has been; from general taxation.

A government ought to be able to budget to provide essential services. That is what the main function of government is. It is what governments are elected to do. Money collected from general taxation is unfairly distributed and too many rich people pay very little taxes. A visit to Bunratty castle and heritage centre will show the contrast between the poor labourer's house and living standard, and the lord in the castle in 16th, 17th, 18th and 19th centuries. The contrast is as great, perhaps even greater today. Compare the multi-millionaire's house in any part of the country with a county council house in some suburban estates in our cities and towns or a small farmers dwelling in many of the counties along the western seaboard.

CHAPTER 5

Ireland's Debt

Ireland's declared richest man, at the height of the Celtic Tiger era, became bankrupt when the house of cards began to crumble. He continues to live in a 14,700 square foot mansion with swimming pool, Jacuzzi, hot tub, a putting green, a sunroom, a conservatory and extensive grounds with mature trees and shrubs. Here we see a Lord of the Manor, modern style. This mansion is in Co, Cavan. Contrast that with a poor farmer in the same county or an unemployed person living with a family of five children in a small country cottage. Compare the rich man's bankrupt position with the poor man's eviction. The poor do not have the money to fight their case or to assert their rights. To add salt to the wounds the Irish taxpayer has to pay for much of the debt incurred, particularly debts relating to the insurance company that was owned by our one-time richest man

The lord of the Manor here is but one of hundreds living in luxury in a broken, depressed, humbled Ireland. Property tax on people's homes was introduced to pay back debts incurred by millionaires, bankers and those who sought to build empires not only in Ireland but they sought to establish Empires abroad too.

Water tax was introduced to force people to pay debts incurred by the government's bailing out of irresponsible bankers, gamblers, speculators, rich developers and big business men.

To repair the damage inflicted by those wealthy, greedy, reckless, speculators people are, not asked but forced to endure greater pain. Spending on hospitals, schools and public services was reduced. Many rural Post Offices, Garda Stations and hospital emergency units have been closed. Hundreds of thousands have been forced to emigrate; many people depressed because of their plight have taken their own lives as pointed out earlier placing Ireland at the top of the suicide scale in Europe. Depression due to a poor future outlook, a deplorable health system, lack of money, evictions, people in serious mortgage arrears, homelessness, unemployment or poor quality jobs, alcohol and drug addiction have all led to greater numbers requiring hospitalization and has contributed to so many unnecessary deaths. Hundreds of families have been evicted from their homes. Along with all the extra taxes that people were forced to pay, public services were also cut or reduced. The introduction of the water tax was the final straw.

Phil Hogan government minister, who introduced the property tax and water charges was becoming very unpopular, so, he was rewarded with a job in Europe with a salary of €336,000 on the first year of taking up office. The person given the job of installing water meters, at an enormous cost to the tax-payer, is a billionaire, one of the wealthiest men in Ireland. He spends a lot of his time in his residence abroad, in Malta, a Mediterranean island where the weather is warm and sunny and skies are blue.

The water tax angered and provoked the Irish people. They accepted the austerity measures heaped on them up to that point. Then they took to the streets in protest with disgust, rage and frustration. They could take no more.

Water is a human right. Where will the money come from to procure a human right? This is nonsense; insulting the people who have already paid for this right through their taxes. The tax system is draconian. It contains an <u>Income tax system</u>, <u>Pay Related Social Insurance</u>, <u>Universal Social Charge</u>, <u>Pension Levies</u>. Then on the take home pay there is; a <u>Property tax</u>, <u>tax on home insurance</u>, <u>tax on house purchase, (stamp duty)</u>, <u>road tax</u>, <u>tax on the</u>

purchase of cars, extremely high taxes on petrol, diesel and alcohol, On top of all that there is a Value Added Tax at a very high rate of 23% on almost all purchases and 13.5% on labour. This leaves even the unemployed and those on other social welfare benefits paying up to 30% of the money they receive back to the state in indirect taxes. It leaves most working people paying between 60% and 70% on what they earn to the government. The people are then asked to pay a water tax.

The war of Independence in America was sparked by a tax on tea. People felt that they were already over-taxed. They rebelled and a bloody war ensued. The French monarchy imposed a salt tax on their people, like the Americans, they were already paying too many taxes. They too rebelled. An even bloodier revolution followed. The guillotine was introduced for the first time. Heads of government rolled in the streets, literally.

The water tax is the last straw for the people of Ireland, but the Irish people want a peaceful revolution. We thought we gained independence when we established our own Parliament (the Dáil) and our own constitution. When we joined the European Economic Community (EEC) and later the Euro zone we shed much of that independence; two treaties in which our own leaders with the backing of the main government parties and opposition forced the people to accept Nice and Lisbon after the people initially rejecting both Treaties.

This caused a further diminution of peoples' independence and sovereignty. Out of the frying pan (under British Colonial rule) and into the fire (under European Imperialism) was where our leaders took us. Either way our independence was always so fragile that we constantly bowed to a foreign power, to America, Britain and Europe depending on them for oil, motor vehicles, information technology and even our daily meals. The emigrants with a proud, industrious hard working Irish origin, a people that played such a big part in building America, Britain, Australia and Canada left behind a dependent, subdued people that now dance to those nation's controlling, capitalist tunes.

THE EUROPEAN VISION

Frederick the Great of Prussia (Germany) tried to control Europe, Maria Antoinette, Bismarck and Hitler continued with that objective; Napoleon and Louis 14th of France had the same ambition. Prior to that, the Roman Empire attempted to control Europe. The Germans tried through war and through diplomatic means.

The present day crusade against Islam is not too unlike the crusades to capture and control Jerusalem and the holy lands in 1095 and again in 1291. Civilization began in Iraq in the valleys of the Euphrates and Tigris rivers. This area was the Garden of Eden. Today Iraq is being destroyed by both Christians and Muslims. Modern day warfare is a mixture of diplomacy, threats and the deployment of weapons of mass destruction mainly in the hands of America, Britain, and Europe.

When we see over-zealous fundamental extremists unleashing their vengeance through terror and barbarous acts of violence against their enemies we are horrified. In the 17th and 18th century's public hangings, beheadings, floggings took place in public places to warn people against acts of treason or other unacceptable crimes. Those inhuman, barbarous acts were attended by thousands of onlookers. The atrocities of the French revolution were a source of entertainment for many people. Earlier in Roman times watching lions killing and devouring people was entertainment for the populous. They even built theatres for such savage performances.

The tower of London, the coliseum in Rome and the holocaust monuments in Germany and Poland all testify to human barbarism. Now, with the development of our civilization when we see beheadings performed by ISIS we are shocked and horrified. We saw in our newspapers recently a picture of onlookers watching a person being burned alive and a young 14-year-old boy enjoying the event. We are numbed, shaken and nauseated as all right-minded, civilized people ought to be by anybody much less a young person taking pleasure in and enjoying such an uncivilized spectacle.

Despite all the advances in information technology, anti-bullying programmes, the Geneva Convention extolling the rights of the human person, and objections to crimes against humanity can we accept or justify the mass genocide of NATO's air to ground bombings on innocent people in Iraq or Israel's treatment of the Palestinians. Thousands are killed every day in such manner but this is referred to as humane killing, collateral damage or merely accidental or necessary deaths to achieve an end. The beheading of one person is so terrible but killing thousands with bombs, missiles and gunfire, leaving children without parents and parents without children in the most horrific manner is taken as less serious, even acceptable, or, worse still as necessary, merely the conesquences of war.

Thinking critically is still not given priority in schools' curricula. Western powers are allowed to develop and store weapons of mass destruction but not the rest of the world. The rationale behind this is the fear that they might use such weapons irresponsibly. There is nothing fair about modern warfare. The end appears to justify the means. When the damage is done when thousands of men, women and children are killed then we hear regrets, apologies and eulogies. We hear phrases like; we are where we are, let's move on; all such talk is of little good to those people who have lost members of their families, friends and companions and have now to pick up the pieces and try to continue with their lives.

World powers extract money from people through taxation to pay for wars, sophisticated missiles to finance banks and keep the super-rich in a luxurious life-style. Taxing tea, salt and water for those purposes has aggravated people and led to war, revolution and protests.

I believe that we should leave the European Union and form a Northern European Alliance with Norway, Sweden, Finland, Denmark, Iceland and Great Britain. We have a lot in common with those countries, much more than we have with Southern Europe. We have similar climate conditions and similar natural resources. We could of course trade and form trade agreements with southern Europe; we could trade water, fish and oil and gas for sunshine, fruit, wines etc. Tourism would be of mutual interest.

CHAPTER 6

Where Will We Get The Money From?

Our health system is in tatters. Our education system is receding. There are over 30 pupils in many classrooms in our main cities and towns in Ireland. It is not possible to cater adequately for such numbers. Many children are not receiving the education required to achieve their potential or a basic standard of education to compete in modern society. Our social welfare system is poor. Hundreds of people are homeless.

Thousands are surviving below an acceptable living standard. The transport service in the main cities of Ireland must be the worst in Europe. Our greatest assets and resources are not progressing but are at stalemate. Money generating industries such as forestry, fishing and farming are vastly under-performing. Smaller industries like Waterford glass and many of the knitting and woollen industries have vanished.

The spire in Dublin's main street is hardly a symbol of Ireland that we want history to portray of our times. The spire (spike) was designed by foreign architects and the stainless steel is not Irish. Industries in the country that are successful are foreign companies using a privileged tax system that was introduced to benefit them.

The first part of the spire was erected in December 2002 after that five additional parts were erected. It was completed on January 21st 2003. I registered my objection to this design as soon as I heard of it and saw what it was to look like. I immediately on the 12th April 1999 nearly 4 years before the first part was erected and before work started on the project, sent a letter of objection to the Tánaiste, the person with responsibility for this spire. I believe that a totally different monument should be considered. My letter was as follows:

12.04.1999

Dear Tánaiste,

I wish to register my strong disapproval to the new millennium spire to be erected in O'Connell Street, Dublin which, I understand, is designed by British architects and will be made from foreign steel.

I enclose a brief sketch of what might be more appropriate to Ireland. I suggest a round tower structure to be made from Wicklow/Dublin granite, representing the monastic and Viking periods of our history. I also suggest that a crystal attachment be included in the construction at the top to capture, magnify and reflect light from the sun at the winter and summer solstices and perhaps the vernal and autumnal equinoxes too. This could be a modern representation of the concepts and magnificent architectural megalithic tomb at New Grange, Knowth and Dowth.

The light from the sun might be directed to shine on the Parnell and O'Connell monuments and on the GPO and James Joyce on other times using artificial reflective methods and light as required to simulate the idea. Artificial light could be used to produce the same effect at midnight on those significant occasions too.

The upper section could represent the modern era with a light on top surrounded by specially commissioned Waterford crystal glass. This would be the symbol of the new millennium, enlightenment.

Around the tower engravings might be inserted on polished granite outlining various events in our history. An elevator with a surrounding spiral stairs should extend up the centre of the tower which could lead to a viewing area and a small coffee bar or restaurant. At the bottom there should be an area paved in Connemara and Kilkenny marble and a surround of Liscannor slate and a low polished seating wall surrounding this area.

A great honour, privilege and responsibility are bestowed on you Tánaiste. I would request that you use your influence to have the erection of the spire reconsidered. I believe it is an insult to the intelligence of the Irish people and particularly the people of Dublin to proceed with a steel spike containing no Irish materials. Our history, our expertise, our native stone, marble and glass industries should be reflected in a structure depicting our times and place in history.

Le gach deá ghuí

Dr. Pat Keogh

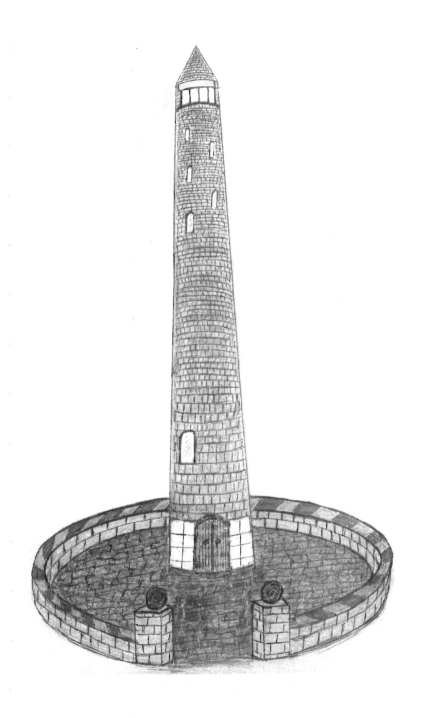

A Round Tower in O'Connell Street instead of the Spire

Total height of tower 70mts, inside diameter 12mts

This round tower should be built with Wicklow/Dublin granite. There might be a patio type area at the bottom of the tower tiled with Liscannor slate surrounded by a low granite wall with seating on top of this wall; seating made from Kilkenny and Connemara marble. There could be two granite pillars at the entrance with a polished granite or bronze ball on top of each pillar.

This tower should have an elevator up the centre, a spiral stairs all around with periodic landings and viewing areas. There could be a full 360 degree viewing area near the top, with a tea/coffee, light refreshments and souvenirs shop in the centre. Above this a Waterford crystal surround (360 degrees) with lights inside that could be changed in colour, depending on seasons or occasions, for example, green for St. Patrick's Day. The outside of the crystal should be directed on the GPO, Parnell and O'Connell at particular times; Winter solstice, vernal and autumnal and summer solstices perhaps.

The doors to the tower should be made from heavy solid Irish Oak and all around the lower part of the tower to door height there should be polished granite slabs embedded in the structure, with inscriptions, symbols or art-work giving a brief history of Ireland --- Starting with the triple spiral from New-grange then going through the bronze and iron ages, some Celtic art or symbols from early Christian or monastic periods, the Viking era, the Normans and up to the 1916 Rising, commemorating the signatories of the proclamation in particular. Our maritime, farming, forestry pursuits could be depicted in art. Irish symbols like the Shamrock and Harp should also be included.

This, I believe would be a reflection of Ireland. It would give Irish people employment, and finances could be generated from an entrance fee to the tower and from refreshments or souvenirs purchased in the restaurant/souvenir shop.

An Alternative for this Commerative Year of The 1916 Rising

A bronze monument of the 7 signatories of the 1916 proclamation would be a fitting tribute to those heroes. Designed, sculptured and erected by Irish people.

Our bright young people are leaving Ireland. They are emigrating, not to Europe but to the English speaking countries of the world; Australia, England, America, New Zealand, and Canada. Ireland has lost its appeal for progressive youths. They have lost interest in their culture, heritage and the Irish language. Those things are seen by many as symbols of depression, suppression and poverty. They are demoralised by the corruption, cronyism and penal tax system that benefits the rich at the expense of the poor. The poor are kept in their place with promises of a bright future. They are talked at in a patronising manner- you are doing great, keep it up. *'But isn't that what you tend to do at times of election'* in the words of one Irish politician.

When the tide rises we will all rise with it but when the boat begins to sink we know who is first to be thrown overboard – the captains go into hiding they call it bankruptcy or liquidation but as the sinking boat begins to rise, after the workers have toiled hard at emptying out the water and salvaging the damaged goods, the steering crew and captain emerge from their hiding places to reap the benefit of a new sailing boat with less on board. They sing Hi Ho! Hi Ho! Off again we go with less on board and the slaves at home let us sail to brighter shores. Hi Ho! Hi Ho! So off again we go.

Where will the money come from? It will undoubtedly come from the pockets of hard working Irish people. The wealthy will continue to enjoy the fruits of the workers' labours. The workers are fobbed off with; *the tide is coming in again.* However, those who know how the tide comes in and goes out at regular intervals are aware that when it goes out again they will once more be left stranded. And so it will continue for the worker. For the super-rich the tide is always in.

In Ireland when our economy collapsed there were two phrases used constantly; 1) <u>*we are where we are,*</u> and 2) <u>*going forward*</u>. Forget the past and let us move on was what those who had perpetrated the crimes, squandered the money and left many people in poverty wanted. Those two phrases were like the chorus of a number one pop song. They were everywhere to be heard and on everybody's lips. It was very clever of the perpetrators to get the people to forget the past, *we are where we are.*

Now, the phrase or question of the moment is: <u>*But where will we get the money from?*</u> Yet some of our wealthiest citizens, one who is worth over 6 billion euro reside outside the state for a sufficient amount of time each year to avoid paying his fair share of taxes here. But this person can and does exercise a great deal of influence over our lives.

Big bankers are back banking and big builders are back building while one family a week is evicted because they are unable to pay their outrageously high home mortgages and rents. Evictions were commonplace when the British Empire controlled Ireland. Now under our own government we are in the clutches of European capitalists. The wealthy in our own land are now doing what the occupiers did in the past.

It is incredible and scandalous that the Irish people have been forced to bail out banks in Ireland and in Europe. We have given somewhere in the region of 60 billion and have received nothing in return. We have per head given more to the European Bank than any other country. That money should have been used to relieve mortgage debt. The banks were greedy as they always have been. They wanted the money to maintain a stranglehold on borrowers. They seek not only, the mortgage repayments but a share in the assets as well. The borrower will not be given the benefit of a write down or even a reduced payment. But, who are the Banks, we might ask? The

Banks are controlled by wealthy capitalists. At the peak of our Celtic Tiger period in Ireland a few wealthy individuals had ambitions to own a bank and a few came close to achieving that ambition.

When banks repossess houses they generally sell that property at a much-reduced rate and they can reduce the mortgage rate to the new owner, but why can't they do this for the existing owner. Surely if a borrower cannot afford to pay the mortgage a solution that is reasonable and affordable should be presented to him/her.

Eviction is an appalling and savage penalty to inflict on a struggling family. It has bitter and horrifying memories for the Irish working people down through the centuries. Evictions, deportations, people exiled followed by plantations of foreigners are memories we would like to forget and many others would like us to forget too, *we are where we are*. The harm that this does to a family is indescribable and irreparable. It leaves a psychological scar on every member of the family, a permanent scar that remains for life often with disastrous and even fatal results.

No family will go through the pain, shame and trauma of allowing their family home to be taken off them through bank repossession if they can prevent it. The family losing a home will still be forced to continue to bail out the very banks that left them without a home. Some families may survive the ignominy and stress of losing their home better than others. They may with great strength, willpower and hard work, start again and get back on their feet. But none will forget the experience of forced eviction. The embarrassment, depression and loss in self-esteem of a person having to go and live with another family member, a neighbour or depend on a charity to help him/her will be engraved in that person's memory forever.

It is incredible to think that a government will see the necessity to save the banks as more important than the lives, health and general well being of its citizens. People's health, children's education and the staff involved in catering for them cannot be compared or modelled on business. Those are not meant to be money making activities.

The Daily mail's COMMENT on January 28th 2016, following the publication of the report on the Banking Inquiry that was set up to examine the causes of the economic crash tells us that the inquiry concluded with: "no single event or decision led to the failure of the banks in the lead-in period to the crisis." People at the top of the banking hierarchy were paid extraordinary high salaries plus allowances. Now, among the inquiry's recommendations is that: "all members of Bank Boards should have requisite financial skills and experience to include banking, risk and governance. It would appear pretty obvious that that should already be the case... but allowing for all this and the report is rather tepid overall – there is a major omission, and it shouts out from the pages: No one is named(blamed) or shamed".

The recommendations are worthless. "Because when a suitable period of time elapses, there really is nothing to stop the whole cycle starting again. Clear and precise apportioning of blame to named individuals was necessary to act as a warning to others. In its absence, the inquiry feels like just another massive waste of money".Now, the phrase, **we are where we are**, is music to those responsible.

CHAPTER 7

A Lateral Thinking Approach

We need a new children's hospital, a much better health system, better care for the elderly, an improved pupil teacher ratio in primary schools, a better pre-school service, restoration on counsellors to second level schools, a better transport system, better use of our waste land, better use of our waters, seas, lakes and rivers, a better developed fishing industry, better tourism, more indigenous industries; wool, leather, cheese, timber and furniture made from our own resources. We need to raise people's pride in our land, our culture, tourism, heritage and our language. We need hope.

Again: where will we get the money from?

Dispense with Presidency for a Children's Hospital!

Gates into Áras aUachtaráin

Áras an Uachtaráin

We should dispense with the presidency. When problems arise and people seek answers the government replies: *Where will the money come from?* One such problem is the need for a new children's hospital. The completion of this hospital has again been delayed till 2019 some children living today will die before then because they will not get the treatment they need.

Replace the presidency with a children's hospital. Which is more important the health and survival of our children, or, the symbolic existence of a monarch living in a mansion? It is pathetic to see people collecting money in buckets on the streets to finance a children's hospital. It is horrific to see sick people flopping sideways on trolleys in hospital corridors, awaiting a bed. It is shameful to see young families evicted from their homes. It is sad to see homeless people begging for crumbs from the rich man's table.

Contrast those images with the image of a president chauffeured as a monarch in a luxurious state car or boarding a chartered flight or government jet to fly to some exotic country. Our president receives a salary of €250,000. The President of America received $400,000 the president of Ireland receives $433,000. The French president has a salary of €180,000 and the Russian president receives €75,000. The Taoiseach is paid more than the Prime Minister of the UK, France, Sweden, Holland, Finland and Spain. The Taoiseach is paid more than the American Secretary of State. Our Tánaiste is also paid more than the Secretary of State in America.

Our president is unable to empathise with the Irish people who voted him into office. The people are protesting in hundreds of thousands as they pay the president's salary and are now asked to pay extra for essential basic necessities. The very high taxes that the people of Ireland pay should provide essential services including a clean water supply.

Our president rose to power through the socialist philosophy. James Connolly founder of socialism in Ireland lived and died in empathy with the plight of the working people. Connolly stood apart, even from the Irish Volunteers of 1916. He thought they were too bourgeois and not sufficiently concerned with Ireland's economic independence. On May 12th 1916 with bullet wounds from gunfire during the Easter rising James Connolly was transported by ambulance from hospital to Kilmainham Gaol where he was carried on a stretcher into the courtyard then tied to a chair, because he was unable to stand due to his feeble state from gunshot wounds. There, tired and weak, he was executed by firing squad. He lived and died for Ireland and particularly for the working people of Ireland.

Contrast that with our president, a Socialist and life-long member of the Labour Party, travelling first class in a luxurious jet to some exotic land, paid for by Irish workers' tax contributions. President Michael D Higgins signed the Water Levy bill into law, without a question, regardless of the wishes of the hundreds of thousands of workers that marched in protest against this unfair and unjust tax; people who would have admired the courage and sacrifice of James Connolly.

Did our president use the socialism of James Connolly, as a platform to achieve the luxury of a king's throne? He walked the Great Wall of China, enjoyed the fruits of South Africa, rubbed shoulders with the upper echelons of the globe while at home people emigrated, were evicted and walked the streets calling for an end to austerity. Many of our homeless people and those living in poverty can identify more with Charles Dickens' times and with Oliver Twist than with Our President: *Please sir! Can In have some more.*

A man in Finglas is ridiculed for using abusive language, expressing annoyance at how the people's president failed to empathise with their plight. That man hadn't time or patience to express his thoughts in an oratorical or poetic fashion. This was his spontaneous, frustrated, un-flavoured rhetoric and we can understand his annoyance. It must be emphasised that it was rude, tasteless and unacceptable. But, how people are being treated is rude, horrible, degrading, tasteless and unacceptable too.

On October 21st 2015 the president on a trip once again to the USA makes a statement that the rest of his presidency will focus and dwell mainly on migration. He is president of Ireland and though it is laudable to speak of the plight of emigrants and he is to be praised for speaking out on their behalf. Many emigrants have suffered beyond human endurance and large numbers have died in pursuit of a place to live in. But there are serious problems in Ireland too. Our president is president of Ireland and not of the world, he should speak out on behalf of our homeless and our ill health system.

Aras an Uachtaráin is valued at €750,000 for property tax purposes, a mansion worth many millions; that tax is paid by the OPW, our money. Who will pay his water bill? We should sell this mansion and close down the office. It stands as a painful reminder of colonial supremacy, times when our people were evicted, times that witnessed the Great Hunger with people suffering and given no medical care. Mr. President! Espouse the ideals of James Connolly who would have stood by people, penalised for the wrongdoings of others, now struggling to achieve better value for their taxes, especially the provision of our own natural resource, water.

People are patronized, belittled, degraded. A referendum was arranged to reduce the age at which one can put him/herself forward for election to the office of presidency from 35 years to 21 years. This was discriminatory in that a person is reckoned to be an adult and has all the rights of an adult at 18 years so why 21 years for this position. The referendum should have been to abolish the presidency.

However, the referendum was not passed, so a person must still be 35 years of age before s/he can become president of Ireland. Yet a person who is 80 or 90 years old is eligible for this office. A 34-year-old person is not eligible to seek the office of president but a 95-year-old person may be president. Surely this leaves the presidency as; not simply a figurehead role, but, a ludicrous, expensive, comical farce. This office has little or no relevance for the vast majority of the people but rather it is a draining office on our struggling people and a limping economy.

Close the Senate for 1000 more nurses in our health system

Savings of over 20 to 25 million a year can be made available by closing the Senate, a totally unnecessary layer of bureaucracy in a small island. A nurses starting pay in Ireland is about €25,000 annually so with between 500 and 1,000 nurses we could open an extra 5000 beds in hospitals around the country one nurse to every six or 8 beds. One of the main stories in the Irish Daily Mail on February 18th 2015 reads: Patient's Lives At Risk; the article continues: "A father of five lying on a trolley for 24 hours said his dignity was left at the door of the emergency department. After 12 hours he had been moved to a corridor with his trolley wedged up against a door. He said that --- his lowest ebb was when a coffin was pushed past him --- people were sick, coughing and all the machines bleeping and you see a coffin passing you" The Mail tells of another patient who had a similar experience. He spent three days in the A&E on February 3rd. He described seeing a crucifix on a coffin going past him; it had a purple sheet draped across it. This for him was the last straw. He packed his bag and attempted to leave without seeing a doctor.

The Irish Nurses and Midwives association INMO say that on the 15th of October 2015 more than 400 beds for the elderly are currently lying idle in community hospitals and public nursing homes. These beds, they say, could accommodate some of the 700 older people languishing in Irish hospitals until more suitable places are found for them.

Basic health-care ought to be a benchmark on how we judge the performance of any government. In an underdeveloped country like Cuba you would not see this type of treatment of its citizens.

Dr. David Hickey, who was a surgeon and Professor in Havana University in Cuba after retiring as Transplant surgeon and kidney specialist in Beaumont hospital, Dublin has given us a great insight into the health system in Cuba

Cuba has no debts. It does not owe any money to any country in the world.

Fidel Castro in 1953 said: *"We were born in a free country that our parents bequeathed to us and the island will first sink into the sea before we consent to be slaves of anyone"*. Ireland's Mission Statement might be the opposite of that.

Cuba values its citizens The American association of world health commented on Cuba's health system in 1993 as follows: *"Cuba has developed an exemplary national health system which provides comprehensive accessible health care to the entire population without charge"*.

Dr. Hickey compares the small budget used for Cuba's health system in comparison with the enormous budget used here.

We need to rid ourselves of the senate or vastly reduce and reform it. The result of the referendum to abolish the senate was too close to be ignored. When the people rejected the Nice European treaty, they were forced to vote again with a vast amount of money spent on marketing. This was Irish and European people's money. European controllers needed Ireland and they persuaded Irish politicians to arrange another Referendum. The government did as they were told. Some people who had voted <u>no</u> the first time were persuaded to change their minds to <u>yes</u> others in frustration or apathy did not vote at all the second time and the political parties got their machines working and brought out their supporters and friends that had not bothered to go out the first time. Consequently the result that Europe wanted was achieved. (Note the Cuban Mission Statement).

A similar operation took place when the Lisbon European Treaty was rejected. 53.4% of those who voted in the Referendum said they were against the Treaty an almost 7% margin between the yes and no votes. People were persuaded to vote a second time, similar to the Nice Treaty, at a huge cost to hard-pressed taxpayers. "Europe for jobs" was the catch phrase used but it really was *jobs for the boys.* 51.7% voted to retain the senate in Ireland, 48.3% wanted to abolish it, a margin if only 3.4%. There was only a 39% turnout to vote on referendum day. Why not repeat the referendum to abolish the senate. No! For that would mean a loss of jobs for the boys. Democracy is a failed system of government. If less than 50% of a population vote at a Referendum or an election how can that represent the majority of the people? Irish citizens who are abroad, most of them because they cannot find employment here, should have a postal vote and if less than 50% cast their vote the election should be declared null and void.

Dispose of Government jets and provide better early childhood schooling

Selling the two government jets The Gulfstream and Lear jet and the annual cost of fuelling, maintaining, piloting and staffing would go a long way to providing better preschool education for children needing it. The government had no hesitation in selling Aer Lingus, but the government jets were retained.

Early intervention for children encountering difficulties at home and in school is so essential. Many learning and socialising problems encountered by a child can be remedied or greatly alleviated if treated early in childhood. The notion of an Irish president, Taoiseach or government minister taking a private jet on a visit to China, South Africa, America etc. is needless, wasteful and appalling since commercial airlines travel so frequently to places all over the world now and if an enormous urgent emergency arises an air corps plane can be used or a private plane can be chartered.

The Minister for Foreign Affairs should manage international diplomatic matters in conjunction with the civil servants that manage each government department. All of the departments have civil servants working for them with a vast amount of accumulated knowledge and information on how best to operate that department. The army should be vastly reduced and the navy and the air corps increased proportionately. Of course we need more than this to get our country back but that might be a start. There are other areas that we ought to streamline or develop in order to provide our people with worthwhile employment. Irish industries should give us pride in our country and improve our economy.

Department of Environment should manage and develop our water system

The company called: "IRISH WATER" is wasting taxpayer's money. <u>Nowhere else is the answer to the question,</u> *where will we get the money from?* More clear, obvious and relevant than disbanding the fiasco *Irish Water.* Is the government involved in treason against its own people? The government has spent over 80 million on consultants' fees and millions on installing meters that will not help the people or make money for the country. They have sent out forms to people costing millions of Euro on printing and postage and thousands on advertising and 40% to 50% of people have not even returned the forms.

No education programme has been put in place and no conservation measure to stop water wastage has been introduced. In fact in many places Irish water workers have caused leaks while installing water meters.

50% less TDs could lower the Pupil Teacher ratio in Primary Schools

We do not need 158 TDs, (members of parliament) 71 would be more than adequate; one per 60,000 people.

Sixty TDs might be elected by the people and eleven chosen for their expertise in particular areas that would be in Ireland's best interest.

The Dáil system is ineffective. The chamber rarely sees all members attend. Imagine running a school, a business or dare I say a hospital where only 20% or 25% of staff turn up for work. All TDs attend only when there is a serious vote that affects the survival of the government and on the first day of a new Dáil. The pairing system encourages abstentions. In parliamentary practice, pairing is an informal arrangement between the government and opposition parties whereby a member of a House of Parliament agrees or is designated by the party **whip** to abstain from voting while a member of the other party needs to be absent from the House due to other commitments, illness, travel problems, etc. A pairing would usually be arranged or approved by the party whips and usually does not apply for critical votes, such as no-confidence votes. This is a clear admission that a no-confidence vote is seen as more important than the day to day business of the Dáil.

The vast majority of the time the chamber has less than 50% of elected members present. The Taoiseach has a very poor attendance record. Questions to the Taoiseach,

Tánaiste or government ministers are rarely answered. Replies to questions are given in a rhetorical, evasive fashion. Waffle and avoidance of answering question has become an art but for many people who see through this tactic, it is nauseating.

Often times a different answer entirely to the question asked is given, the answer given is the answer to a question that the leader or minister would like to have been asked. They answer in a cynical, comical, pedantic or farcical manner, often times finishing with a grin – game playing (See how I answered or didn't answer that one is portrayed in the smirk). This system of government is a waste of taxpayers' money. People are entitled to hear honest answers to questions put by an opposition TD, a backbench TD or to a parliamentary question initiated by an individual, organisation or a concerned group of people and asked by a TD on his/her or their behalf.

Disband Health Service Executive for better health system

We do not need two overlapping systems like the Department of Health and the HSE. The Health Service Executive (HSE) is over bureaucratic, with too many administrators producing little or no benefit to the patient. It is wasteful to have a whole civil service to oversee the work of the Department of Health and another layer in the Health Service Executive. The Department of Health should be the body that is responsible for our health system under the direction of a Minister for Health. The Department of Health could be sub-divided into community care, hospitals and clinics. The old Health Boards were better than the Health Service Executive. Top administrators are paid too much while doctors are over worked. Many hospitals do not have enough doctors, nurses and carers.

The chief director of the Health Service Executive (HSE) is paid €185,300 plus expenses. Other administrators in the system are paid high salaries that would be better used in providing more nurses and care assistants.

Disband County councils and departments of environment; replace with a smaller more efficient local government system

We do not need the County Councils and the Department of the Environment. County Councillors cost the tax payers an enormous amount of money including the financing of council chambers, election costs, junkets, daily expenses lighting, heating, cleaning and maintenance of premises etc. County Council Chambers are idle for most of the time.

Reduce the army, increase the Navy and retain the air-corps

The state has remained neutral in war times and has a long-standing policy of non-aggression in armed conflicts. The Defence Force has three branches: the Army, Naval Services and Air Corps. The Irish Army has 8,500 active personnel and 4,000 in the Army Reserve.

We need to increase the naval service, retain or increase the air-corps and reduce the army. The Navy could do the same work as the army in times of crises; floods, snowfall periods etc.

Our minister for Defence has suggested that we should create our own fleet of fighter jets capable of firing air to air missiles. This proposal has been suggested in the light of an alleged Russian airplane crossing over our air space. The cost to the Irish taxpayer would be in the region of €30 million per air craft and we would require a minimum of six. Then we have the cost of training pilots to operate such aircraft, this would cost millions of euro – *Where will we get the money from?* This question is not asked by government in this situation. But we, the people might ask it. Many homeless people are kept in hotels because the government has not built houses for them. It is reported in the news 19th October 2015 that one lady and her daughter is being kept in a hotel at a cost of nearly €5000 per month, roughly the cost of two house mortgages at the same time the newly announced extra year's child care/pre schooling announced in last week's budget in now being delayed.

The Air Corps is the smallest branch of the Defence Forces, with approximately 939 personnel.

The role of the Air Corps is to defend Irish airspace, however, its capability to do so is so limited that it is practicable negligible. The Air Corps also provides non-military air services such as air ambulances, VIP transport, and search and rescue (in support of Coast Guard search and rescue efforts).

Naval Service

The Naval Service has about 1,144 personnel, and is responsible for guarding Irish territorial waters as well as the Irish Conservation Box a large area of sea in which fishing is restricted to preserve fish numbers. The Naval

Service is tasked with enforcing this EU protected area and so serves the EU as well as Ireland. The Naval Service, together with the Air Corps and Customs, has intercepted a number of vessels carrying narcotics to and from Ireland.

The Naval Service has eight offshore patrol vessels with highly trained armed boarding parties that can seize a vessel if necessary, and a special Naval Service Diving Section. While the Naval Service does not have any heavy warships, all of the naval vessels have enough firepower to enforce their policing roles.

The Navy gave very valuable assistance in 2015 to rescuing migrants from the waters of the Mediterranean Sea as they escaped from war-torn regions in Africa and the Middle East. The army's main function seems to involve peace keeping abroad. It must be acknowledged that they were used to great effect during the floods in December 2015 and January 2016. They could be more visible and employed on much more home missions such as involvement in forestry and assisting local government with emergencies

CHAPTER 8

Forestry - Fishing - Tourism

Forestry needs to be developed and expanded in the enormous amount of mountain areas and cut away bog lands in Ireland. In years to come future generations would reap the benefits and they would secure a valuable source of income from such a development. The amount of worthwhile employment that could be generated initially through draining the lands and other preparatory work even before planting starts would be enormous. This would not be employment for the sake of employment or just to massage the figures and keep them right for government propaganda.

This industry would provide continuous, sustainable employment, as thinning and constant maintenance work would be necessary. Replanting trees and replacing diseased or decaying trees would always be necessary. Harvesting mature trees and the industries that could be created from our homegrown timber; larch, beech, birch, ash, spruce and the slower growing oak and many varieties of conifers would be continuous.

The Fishing Industry ought to be a natural industry in Ireland. This involves boat making and boat and ship maintenance as well as making and maintaining all types of fishing equipment. Fish processing, tinning, freezing and deliveries of fresh fish to hotels and restaurants would provide ongoing, sustainable employment. Spawning beds need to be developed and the subsequent stocking of our rivers and lakes for pleasure fishing and fresh water fish for export would be sustainable Industries too. Research on how best to operate and oversee this industry should be a faculty in third level colleges. There should be a Department of fisheries and a Department of forestry with well-trained civil servants, college graduates and experts to cater for these industries. Keeping rivers and lakes pollution free would need to be stressed and encouraged in our education system. Policing is also necessary.

Tourism is vastly underdeveloped in Ireland. We need to promote the uniqueness of our culture: heritage, dance, games and language.

Our many archaeological sites require development and promotion. Marketing firms, historians, archaeologists and architects are needed to develop this industry.

Aer Lingus has been sold. This was not in Ireland's best interest. A country without its own airline has little to identify it and leaves it soul-less. Irish people landing in foreign airports always looked out for Aer Lingus planes when they landed. This was a great advertising symbol.

We have so many sites, ruins, buildings and items to display to visitors: New grange, Knowth and Dowth, The book of Kells, Innis Fallen island and the annals of Innis Fallen, Glendalough's round tower and monastic site and graveyard, Clonmacnoise round tower and monastic site, Waterford Viking town and Reginald's tower, the Giants Causeway, the Ceide Fields, the Battle of the Boyne Interpretative Centre, Tara site of the early high kings of Ireland, the Aran islands, The Rock of Cashel, The Claddagh; Ail wee caves in Co. Clare, Mitchelstown caves in Cork and Crag caves in Castleisland, Co. Kerry. Fore monastic settlement in County Westmeath, Trim Norman castle in Co Meath, Killarney, Dingle, Connemara, West Cork, Kilkenny castle, Dublin Castle, Christchurch and St. Patrick' Cathedrals, St. Michan's Church, Glasnevin cemetery and in Wicklow Powers Court house and gardens to name but a few.

Book of Kells

Conclusion

The most important and fairest way to get the money required to run the country efficiently is to make the rich pay their fair share of taxes. Rich people should not be allowed to evade paying tax by living outside the state for a specified number of days or months each year. They should pay tax on the income or profits they make in this county. In a report in The Irish Daily Mail by Paul Caffrey on May 13th 2015 we read:" The Malta-based billionaire yesterday launched a bid to stop RTE broadcasting a report describing him as a one-time 'major debtor' to the former Anglo Irish bank. --- the broadcast was postponed after Denis O'Brien's lawyers sought an injunction against the national broadcaster."

Where will we get the money from to pay for water? We pay already through our general taxation system. We pay a second time in schools, colleges, hospitals, industries and businesses pay water rates. Children and their parents pay as schools and colleges are funded by the tax payer, the sick in hospitals pay and since all businesses and manufacturing industries pay for water that cost is added to the items produced by those industries or businesses. Surely it is unreasonable, aggravating and provocative to ask people to pay again.

My main objection to paying directly for water is that water is a human right and it is Ireland's greatest natural resource and should be financed from general taxation as it always has been.

I will conclude with the words of Luke Kelly's monologue: *For What Died The Sons Of Roisin* (Luke was s singer with the folk and Ballad group; The Dubliners): Irish governments seem to be intent on privatising and selling out all our services. What will be next Bord na Mona, Transport service (a start has been made here with the sale of flights to the Aaron islands to a private company) our forests, our water, and perhaps the entire tourist industry.

For What Died The Sons Of Roisin
Luke Kelly (The Dubliners)

For What Died the Sons of Róisín, was it fame?
For What Died the Sons of Róisín, was it fame?
For what flowed Irelands blood in rivers,
That began when Brian chased the Dane,
And did not cease nor has not ceased,
With the brave sons of '16,
For what died the sons of Róisín, was it fame?

For What Died the Sons of Róisín, was it greed?
For What Died the Sons of Róisín, was it greed?
Was it greed that drove Wolfe Tone to a pauper's death in a cell of cold wet stone?
Will German, French or Dutch inscribe the epitaph of Emmet?
When we have sold enough of Ireland to be but strangers in it
For What Died the Sons of Róisín, was it greed?

To whom do we owe our allegiance today?
To whom do we owe our allegiance today?
To those brave men who fought and died that Róisín live again with pride?
Her sons at home to work and sing,
Her youth to dance and make her valleys ring,
Or the faceless men who for Mark and Dollar,
Betray her to the highest bidder,
To whom do we owe our allegiance today?
For what suffer our patriots today?
For what suffer our patriots today?
They have a language problem, so they say,
How to write "No Trespass" must grieve their heart full sore,
We got rid of one strange language now we are faced with many, many more,
For what suffer our patriots today?